Routledge Revivals

Sensationalism and Scientific Explanation

Sensationalism and Scientific Explanation is a critical examination of the view that scientific statements can be understood only in terms of basic 'atoms' of experience, also called 'sensations'.

Presenting different extremes of this view, the book considers whether it can provide an adequate account of science as we find it. It explores in detail the sensationalist account of science set out by Ernst Mach in relation to various aspects of scientific investigation and theorizing, and puts forward an argument for the 'inherent weakness of sensationalism'.

Sensationalism and Scientific Explanation will appeal to those with an interest in the history and philosophy of science.

Sensationalism and Scientific Explanation

By Peter Alexander

First published in 1963
by Routledge & Kegan Paul Ltd.

This edition first published in 2020 by Routledge
4 Park Square, Milton Park, Abingdon, Oxon OX14 4RN

and by Routledge
605 Third Avenue, New York, NY 10017

Routledge is an imprint of the Taylor & Francis Group, an informa business

© Peter Alexander 1963

All rights reserved. No part of this book may be reprinted or reproduced or utilised in any form or by any electronic, mechanical, or other means, now known or hereafter invented, including photocopying and recording, or in any information storage or retrieval system, without permission in writing from the publishers.

Publisher's Note
The publisher has gone to great lengths to ensure the quality of this reprint but points out that some imperfections in the original copies may be apparent.

Disclaimer
The publisher has made every effort to trace copyright holders and welcomes correspondence from those they have been unable to contact.

A Library of Congress record exists under LCCN: 63002831

ISBN 13: 978-0-367-61090-6 (hbk)
ISBN 13: 978-1-003-10408-7 (ebk)
ISBN 13: 978-0-367-61089-0 (pbk)

For Product Safety Concerns and Information please contact our EU representative GPSR@taylorandfrancis.com Taylor & Francis Verlag GmbH, Kaufingerstraße 24, 80331 München, Germany

Printed and bound by CPI Group (UK) Ltd, Croydon, CR0 4YY
08/06/2025
01896991-0002